W9-BYY-089

Geography Zone: Landforms™

BAYS

Emma Carlson Berne

PowerKiDS press™

New York

Published in 2008 by The Rosen Publishing Group, Inc.
29 East 21st Street, New York, NY 10010

First Edition

Editor: Joanne Randolph
Book Design: Julio Gil
Photo Researcher: Jessica Gerweck

Photo Credits: Cover © iStockphoto.com/Robert Simon; p. 5 © iStockphoto.com/Todd Taulman; p. 7 © iStockphoto.com/Ann Steer; p. 9 © iStockphoto.com/S. Greg Panosian; p. 11 © iStockphoto.com/Maciej Laska; p. 13 © iStockphoto.com/Ina Peters; pp. 15, 17 Shutterstock.com; p. 19 © iStockphoto.com/Kenneth C. Zirkel; p. 21 © iStockphoto.com/Dennis Tangney.

Library of Congress Cataloging-in-Publication Data

Berne, Emma Carlson.
   Bays / Emma Carlson Berne. — 1st ed.
      p. cm. — (Geography zone : Landforms)
Includes index.
   ISBN 978-1-4042-4206-7 (library binding)
   1. Bays—Juvenile literature. I. Title.
   GB454.B2B47 2008
   551.46'18—dc22
                                                2007034532

Manufactured in the United States of America

Contents

A bay is a body of water that has land on three sides. Bays are on the coasts of oceans. Bays are usually in the shape of a **semicircle**. Sometimes, they are almost a full circle.

Bays are found on the coasts of every **continent** of the world. Some bays are very small, only a few hundred yards (m). Other bays are very big. Big bays can be several hundred miles (km) across.

Hudson Bay is a very big bay in Canada. It is 319,000 square miles (826,206 sq km).

People enjoy visiting beaches along bays. This is Horseshoe Bay and the beach alongside it on the island of Bermuda, off America's East Coast.

Bays are made by the ocean over a long time. The ocean waves wash against the coast over and over. After a long time, the ocean wears away a shape in the coastline. This shape is the bay.

On either side of bays are headlands. Headlands are the lands that are left sticking out into the ocean next to the bay. Some big headlands are called **capes**.

At the southern tip of Africa is a headland called the Cape of Good Hope. The Cape of Good Hope is famous among sailors for its big waves and strong winds.

This is the Cape of Good Hope and part of False Bay. The cape is part of Cape Peninsula. A peninsula is land with water on three sides.

Ocean waves **erode** the rock of the coastline to make the bay. Coastlines are made of different kinds of rock. Some of the rock is hard and some is soft.

The hard and soft rock sometimes lie right next to each other. In the places where the rock is soft, the ocean waves can erode the rock more quickly. Over time, the waves wash the rock away and make the shape of the bay.

Next to the soft rock is hard rock. The ocean cannot wash this rock away as easily. The hard rock remains, sticking out into the sea, making a headland.

Here the waves crash against the rocks along California's coast. Over time, the waves wear away the rock and cut a bay into the coastline.

Some headlands have sea caves in them. Sea caves are made when the ocean water splashes into a crack, or small opening, in a headland. The water has sand and small stones in it. After a long time, the water, stones, and sand erode the rock. The crack gets wider. Finally, a hollow place appears. This is the cave.

If the water, gravel, and stones keep eroding the rock, the cave gets deeper and deeper. The back of the cave can go right through to the other side of the headland. This makes an **archway**. Some archways are big enough for a boat to sail through.

These arches are found at the Blue Caves on the island of Zákinthos, in Greece. This island has many small bays and caves along its coast.

Sometimes, the waves keep wearing away at the rock of a cave. The archway gets thinner and thinner. Eventually, it gets so thin that the top falls in. A tall tower of rock is left sticking out of the ocean. This is called a stack. Birds like to make their nests on stacks, out of the way of **predators**.

After a long time, the stack wears down, too. The waves wash over it again and again, so that only a small, rounded stump is left sticking out of the ocean. The stump is usually underwater when the tide comes in.

It is said that pirates, or sailors who rob towns and boats, used Italy's bay of Scopello as a base. The pirates hid their ships behind these stacks.

Many animals like bays. They like the calm water that is protected from big waves. Crabs, sea slugs, worms, and shellfish live in bays. They live down near the bottom of the bays, among the green and yellow seaweed.

Fish like cod, haddock, and bass swim in the deep water of bays. Eels hide in underwater holes in the rock of the shore. Jellyfish float along in the water, too. Near the top of the water, seals and dolphins swim. They come to the bay to play and look for fish to eat.

Many sea plants do well in bays, too. Some of these include seaweed and sea grasses.

Crabs, like this one, like to make their homes in or near bays.
Some crabs live on the floor of the bay and others live on the shore.

Sometimes, grass grows on the top of headlands. These grasses wave in the wind from the ocean. Trees, like cypress trees, can grow on headlands, too. Headlands are rocky, though. Sometimes, there is not enough dirt for plants to grow very well.

Sea birds like to live on headlands. Seagulls, puffins, and terns build their nests on the rocks that overlook the ocean. Down near the water, barnacles and other **crustaceans** hang on to the rocks where the ocean waves can splash them. Otters sometimes lie on the rocks near the water and **bask** in the sun.

Puffins spend nearly their whole lives at sea. They come on land only between April and mid-August to nest on rocky headlands and coasts.

Massachusetts has a famous cape called Cape Cod. Glaciers formed Cape Cod between 16,000 and 20,000 years ago. The cape is 65 miles (105 km) long and only 1 mile (2 km) wide in some spots! It looks a little like a curled arm with a fist at the end. Next to Cape Cod is a large bay, called Cape Cod Bay.

Long ago, the only people who lived on Cape Cod were a Native American group called the Wampanoag. In 1620, the Pilgrims landed on Cape Cod and stayed a little while before they sailed to Plymouth.

This lighthouse sits at the very tip of Cape Cod. Cape Cod's beautiful beaches and seashore towns bring in many visitors each summer.

Today, many people live on Cape Cod. Villages are scattered up and down the coast. Many **tourists** come to Cape Cod every summer to swim in the ocean and lie on the beaches. The Cape Cod National Seashore is one of the most-visited places on Cape Cod. It includes many beaches and protected places for wildlife.

Fishing is an important **industry** on Cape Cod. Every day, fishing boats pull out of the **harbors** and sail off to fish for lobster, crab, and cod.

Farmers grow cranberries and asparagus on Cape Cod. The soil is so sandy that not many other crops will grow there.

Cranberries are being harvested, or picked, here. Of the 1,000 cranberry growers in the United States, 500 of them are in Massachusetts.

Bays and headlands are important to people all over the world. They provide protection from the rough ocean waves and deep water. People can swim and fish in the calm, peaceful water of bays. They can sail boats that would be too little for the wide open ocean.

Bays make excellent harbors for ships. Sailors can **anchor** their ships in bays to keep them safe from storms. Bays can be shortcuts for carrying things people need. Ships with goods can sail from one side of a bay to another in a short time.

Glossary

anchor (AN-ker) To hold a ship in place using a piece of metal tied to a ship and thrown overboard.

archway (AHRCH-way) An opening that is curved, or rounded, at the top.

bask (BASK) To lie in the sun.

capes (KAYPS) Points of land that stick out into the water.

continent (KON-tuh-nent) One of Earth's seven large landmasses.

crustaceans (krus-TAY-shunz) Animals that have no backbone and have a hard shell, and live mostly in water.

erode (ih-ROHD) To wear away slowly.

harbors (HAR-borz) Safe bodies of water where ships anchor.

industry (IN-dus-tree) A business in which many people work and make money producing something.

predators (PREH-duh-terz) Animals that kill other animals for food.

semicircle (SEH-mee-sur-kul) Half of a circle.

tourists (TUR-ists) People visiting a place where they do not live.

Index

A
archway(s), 10, 12

C
cape(s), 6, 18, 20
Cape Cod Bay, 18
Cape of Good Hope, 6
cave, 10, 12
coast(s), 4, 6, 20
continent, 4

H
harbors, 20, 22

headland(s), 6, 8, 10, 16, 22
headland(s), 6, 8, 10,
  16, 22
Hudson Bay, 4

O
ocean(s), 4, 6, 8, 12,
  16, 20, 22

P
predators, 12

R
rock(s), 8, 10, 12, 14, 16

S
sailors, 6, 22
sand, 10
sea, 8
stones, 10

T
tourists, 20

W
water, 4, 10, 14, 16, 22
waves, 6, 8, 12, 14, 16,
  22

Web Sites

Due to the changing nature of Internet links, PowerKids Press has developed an online list of Web sites related to the subject of this book. This site is updated regularly. Please use this link to access the list:
www.powerkidslinks.com/gzone/bay/